Volcanoes in Mexico

by Barbara Wood

This is a mountain.
It is not a volcano.
But some mountains are volcanoes.

This mountain is a volcano.
It has an opening in it.

This mountain is a volcano.
It has snow on its top.

This mountain is a volcano.
It can erupt with lava or steam.

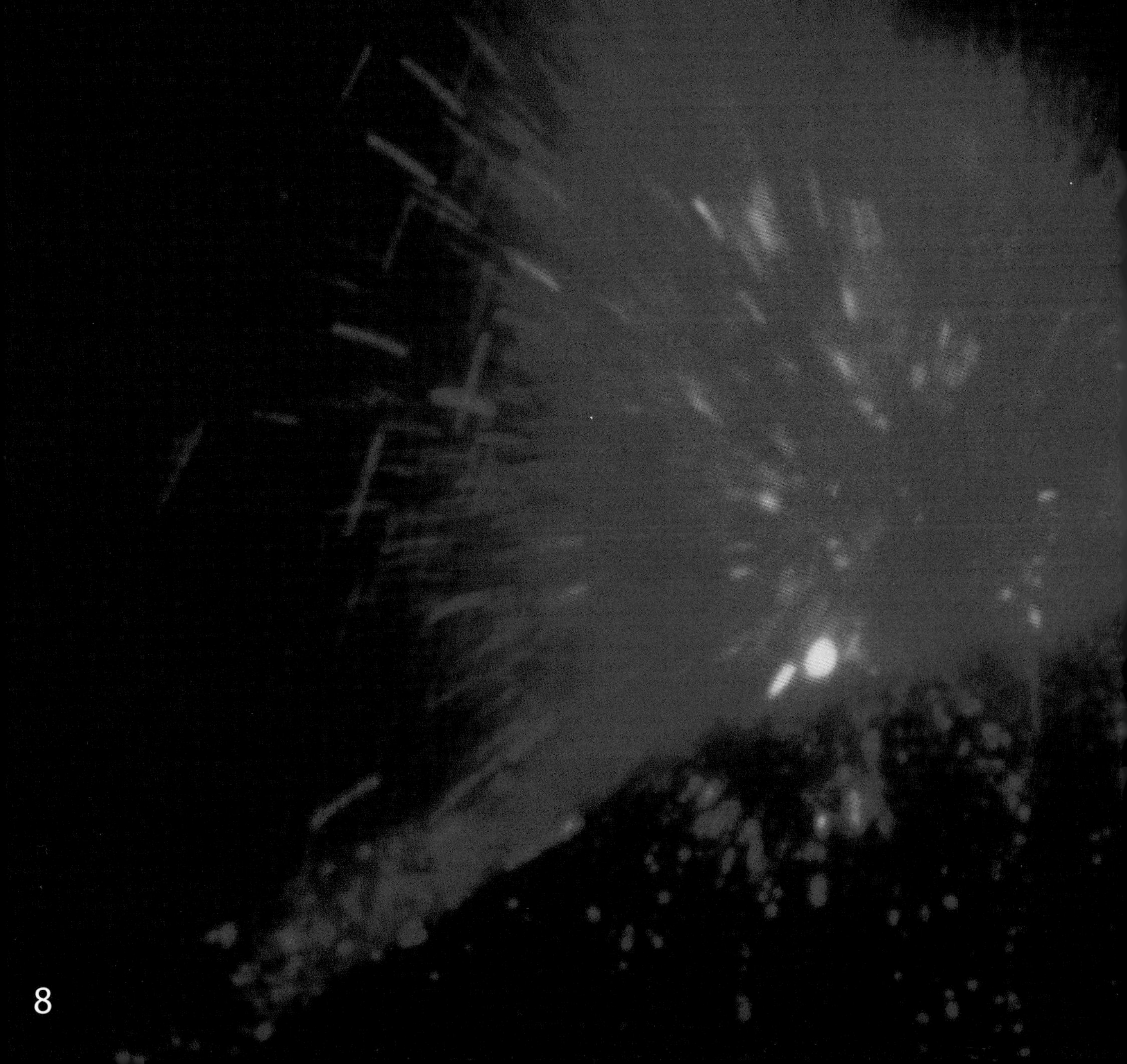

This mountain is a volcano.
It can erupt with ash or lava.

lava

Did You Know?

An earthquake cracked this field.
Smoke came out of the crack.
Then a volcano erupted!

Michoacán, Mexico, 1943